woolly wisdom

For Tyla

Thanks to Renée Lang, whose idea it was to create this book

First published in 2004 by New Holland Publishers (NZ) Ltd
Auckland • Sydney • London • Cape Town

218 Lake Road, Northcote, Auckland, New Zealand
14 Aquatic Drive, Frenchs Forest, NSW 2086, Australia
86–88 Edgware Road, London W2 2EA, United Kingdom
80 McKenzie Street, Cape Town 8001, South Africa

www.newhollandpublishers.co.nz

ISBN: 1 86966 063 3

Publishing manager: Renée Lang
Project editor: Fionna Campbell
Design: Trevor Newman

A catalogue record for this book is available from the National Library of New Zealand

10 9 8 7 6 5 4 3 2 1

Colour reproduction by SC (Sang Choy) International Pte Ltd, Singapore
Printed in China through Phoenix Offset, Hong Kong

woolly wisdom

DON DONOVAN

Groups tend to agree on courses of action which, as individuals, they know are stupid.

ANONYMOUS

The secret of staying young is to live honestly, eat slowly, and lie about your age.

LUCILLE BALL

Generally speaking, it's a matter of only mild intellectual interest to me whether the earth goes around the sun or the sun goes around the earth.

EDWARD ABBEY

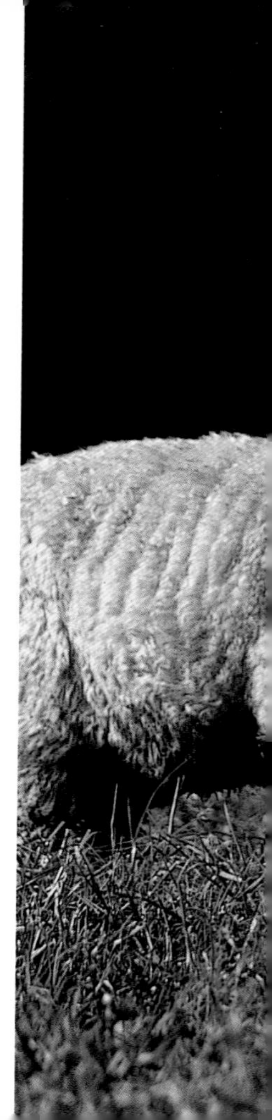

When people are free to do as they please, they usually imitate each other.

ERIC HOFFER

Be content with your lot; one cannot be first in everything.

AESOP

Weeds are flowers too, once you get to know them.

A.A.MILNE (Eeyore from *Winnie the Pooh*)

TR951

We know what happens to people who stand in the middle of the road. They get run over.

ANEURIN BEVAN

Remember, today is the tomorrow

you worried about yesterday. DALE CARNEGIE

No wonder I'm unhappy –
My twin forgot my birthday.

JERRY DENNIS

I think animal testing is a terrible idea; they get all nervous and give the wrong answers.

JOSEPH BLOSEPHINA

Men are like tile floors, lay them down right the first time and you can walk all over them for the rest of your life.

LISA TARBOX

Food is an important part of a balanced diet.

FRAN LEBOWITZ

When you come to a fork in the road

... take it. YOGI BERRA

There aren't any embarrassing questions – only embarrassing answers.

CARL ROWAN

Oh, why does the wind blow upon me so wild? Is it because I'm nobody's child?

PHILA HENRIETTA CASE

No one thinks of winter when the grass is green!

RUDYARD KIPLING

**Only good girls keep diaries.
Bad girls don't have the time.**

TALLULAH BANKHEAD

He that fights and runs away
May live to fight another day.

ANONYMOUS

**The way I see it,
if you want the rainbow,
you gotta put up
with the rain.**

DOLLY PARTON

If all the girls who attended the

Yale prom were laid end to end,
I wouldn't be a bit surprised. DOROTHY PARKER

Never do anything standing that you can do sitting, or anything sitting that you can do lying down.

CHINESE PROVERB

When I was just a little girl
I asked my mother,
what will I be?
Will I be pretty, will I be rich?
Here's what she said to me.

Que Sera, Sera,
Whatever will be, will be
The future's not ours to see
Que Sera, Sera
What will be, will be.

LIVINGSTON & EVANS

Never assume,
for it makes an ASS
out of U and ME.

ANONYMOUS

So, where's the Cannes Film Festival being held this year?

CHRISTINA AGUILERA

What is a friend? A single soul dwelling in two bodies.

ARISTOTLE

Women are like stars,
there are millions of them out there,
but only one can make your
dreams come true.

ANONYMOUS

Before God we are all equally wise

– and equally foolish. ALBERT EINSTEIN

4830

One main factor in the upward trend of animal life has been the power of wandering.

ALFRED NORTH WHITEHEAD

'Mid pleasures and palaces
though we may roam.
Be it ever so humble,
there's no place like home.

J.H.PAYNE

I may not be totally perfect, but parts of me are excellent.

ASHLEIGH BRILLIANT

'Afoot and light-hearted

**I take to the open road
Healthy, free, the world before me.'**

WALT WHITMAN

Contraceptives should be used on every conceivable occasion.

SPIKE MILLIGAN. THE LAST GOON SHOW OF ALL

Don't ever take a fence down until you know why it was put up.

ROBERT FROST

The House of Commons starts its proceedings with a prayer. The chaplain looks at the assembled members with their varied intelligence and then prays for the country.

LORD DENNING

As you get older three things happen. The first is your memory goes, and I can't remember the other two...

SIR NORMAN WISDOM

Beware of all enterprises that require new clothes.

HENRY DAVID THOREAU

You can complain because roses have thorns, or you can rejoice because thorns have roses.

ZIGGY

Q: 'What do you call an Australian with a hundred girlfriends?'
A: 'A shepherd.'

AUSTRALASIAN JOKE